THE LITTLE BOOK OF
ELEMENTS

Published by OH!
20 Mortimer Street
London W1T 3JW

Text © Jack Challoner 2020
Design © 2020 OH!

ISBN 978-1-91161-057-1

Editorial: Jack Challoner, Lesley Levene
Project manager: Russell Porter
Design: Andy Jones
Production: Rachel Burgess

A CIP catalogue record for this book is available from the British Library

Printed in Dubai

10 9 8 7 6 5 4 3 2 1

THE LITTLE BOOK OF
ELEMENTS

JACK CHALLONER

CONTENTS

PREFACE

Most familiar substances are not elements. Wood, steel, air, salt, concrete, water, plastics, glass – these are all mixtures or compounds, made up of more than one element. We do encounter elements in a fairly pure state in our everyday life. Gold and silver are good examples, as are copper (pipes), iron (railings), aluminium (foil) and carbon (as diamond). Some other elements are familiar simply because they are so important or commonplace. Oxygen, nitrogen, chlorine, calcium, sodium, lead – these are all examples of such elements.

This book will explore the history of each of the elements in the periodic table (overleaf), including the derivation of the elements' names. The properties of each element will also be explored, including its chemical behaviours – in other words, how its atoms interact with atoms of other elements. So, for each element, we will also look at some important compounds and mixtures that contain it.

Jack Challoner

THE PERIODIC TABLE

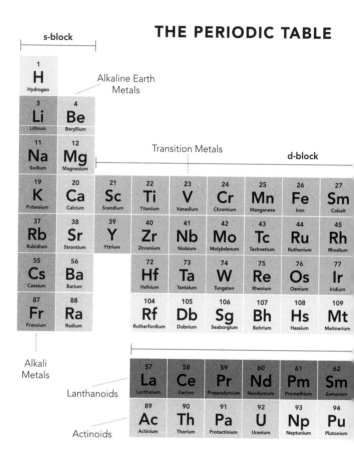

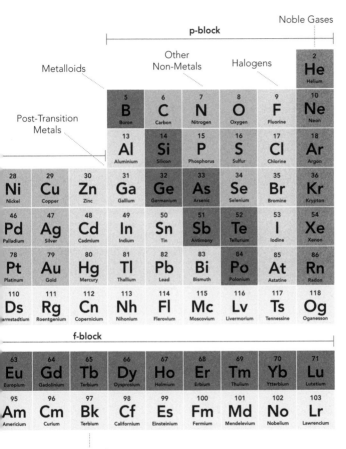

Noble Gases

p-block

Other
Non-Metals

Halogens

Metalloids

2
He
Helium

Post-Transition
Metals

5	6	7	8	9	10
B	**C**	**N**	**O**	**F**	**Ne**
Boron	Carbon	Nitrogen	Oxygen	Fluorine	Neon

13	14	15	16	17	18
Al	**Si**	**P**	**S**	**Cl**	**Ar**
Aluminium	Silicon	Phosphorus	Sulfur	Chlorine	Argon

28	29	30	31	32	33	34	35	36
Ni	**Cu**	**Zn**	**Ga**	**Ge**	**As**	**Se**	**Br**	**Kr**
Nickel	Copper	Zinc	Gallium	Germanium	Arsenic	Selenium	Bromine	Krypton

46	47	48	49	50	51	52	53	54
Pd	**Ag**	**Cd**	**In**	**Sn**	**Sb**	**Te**	**I**	**Xe**
Palladium	Silver	Cadmium	Indium	Tin	Antimony	Tellurium	Iodine	Xenon

78	79	80	81	82	83	84	85	86
Pt	**Au**	**Hg**	**Tl**	**Pb**	**Bi**	**Po**	**At**	**Rn**
Platinum	Gold	Mercury	Thallium	Lead	Bismuth	Polonium	Astatine	Radon

110	111	112	113	114	115	116	117	118
Ds	**Rg**	**Cn**	**Nh**	**Fl**	**Mc**	**Lv**	**Ts**	**Og**
armstadtium	Roentgenium	Copernicium	Nihonium	Flerovium	Moscovium	Livermorium	Tennessine	Oganesson

f-block

63	64	65	66	67	68	69	70	71
Eu	**Gd**	**Tb**	**Dy**	**Ho**	**Er**	**Tm**	**Yb**	**Lu**
Europium	Gadolinium	Terbium	Dysprosium	Holmium	Erbium	Thulium	Ytterbium	Lutetium

95	96	97	98	99	100	101	102	103
Am	**Cm**	**Bk**	**Cf**	**Es**	**Fm**	**Md**	**No**	**Lr**
Americium	Curium	Terbium	Californium	Einsteinium	Fermium	Mendelevium	Nobelium	Lawrencium

Transuranium Elements

CHAPTER

1

AN INTRODUCTION TO ELEMENTS

PLEASE READ!

This introduction contains crucial information that will enable you to understand the organization of this book and all the details it contains, including the data files that accompany each element's entry.

PROTONS, NEUTRONS AND ELECTRONS

An element is a substance with only one kind of atom. An atom has a diameter in the order of one ten-millionth of a millimetre (0.0000001 mm, 0.000000004 inches). Its mass is concentrated in a heavy central part, the nucleus, made of particles called protons and neutrons – the number of protons present in the nucleus determines which element the atom belongs to. Much lighter particles, called electrons, surround the nucleus.

Protons carry positive electric charge; electrons carry a corresponding amount of negative electric charge. Scale them up in your imagination, so that they are little electrically charged balls you can hold in your hand, and you would feel them pulling towards each other because of their mutual electrostatic attraction. Neutrons are neutral: they carry no electric charge. Hold a scaled-up one of these in your hand, and you will see that it is not attracted towards the proton or the electron.

BUILDING ATOMS

With these imaginary, scaled-up particles, we can build an atom of the simplest and lightest element, hydrogen. For the nucleus of your hydrogen atom, you just need a single, naked proton. To that, you will need to add your electron – by definition, an atom has equal numbers of protons and electrons, so that it has no charge overall. Hold the electron at some distance from the proton and the two particles will attract, as before. The force of attraction means that the electron has potential energy, just as a stone has potential energy if you hold it some distance above the ground.

Let go of the electron and it will "fall" towards the proton, losing potential energy. Thanks to the weird rules of quantum physics, which govern the behaviour of subatomic particles, the electron falls in definite steps, rather than a continuous descent. It also stops short of crashing into the proton, settling instead into an orbit around it. It is now in its lowest possible energy state, n=1. The steps above it, at which the electron paused on its descent, are at n=2, 3 and so on.

In the unfamiliar world of quantum physics, particles can be in more than one place at the same time, and as well as being tiny, well-defined objects, they exist as spread-out waves. So at the same time as being a well-defined particle, your electron is also a three-dimensional "stationary wave". And so, your electron appears as a fuzzy sphere surrounding the nucleus, called an orbital. The chemical properties of elements are determined mostly by the arrangement of electrons in orbitals around the nucleus.

Illustration of an orbital, the region in which electrons can exist.

ATOMIC NUMBER

Now move the electron away, leaving the naked proton again. To make the next element, with atomic number 2, you will have to add another proton to your nucleus. But all protons carry positive charge, so they strongly repel each other – and the closer you move the protons together, the more strongly they push apart. Try adding a neutron instead – there is nothing stopping you this time, because the neutron has no electric charge.

As you bring the neutron very close, you suddenly notice an incredibly strong force of attraction, pulling the neutron and proton together. This is the strong nuclear force – it is so strong that you will now have trouble pulling the proton and neutron apart. This force only operates over a very short range.

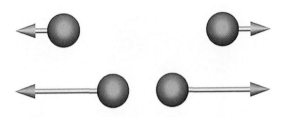

The repulsive (electrostatic) force between two protons is stronger the closer they are.

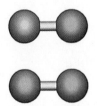

The attractive force between protons and neutrons has a very limited range.

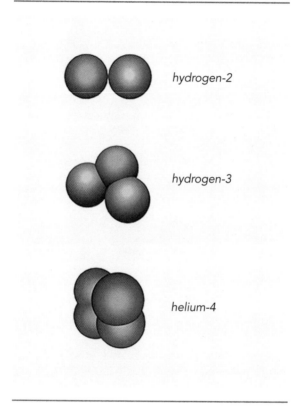

hydrogen-2

hydrogen-3

helium-4

You now have a nucleus consisting of one proton (1p) and one neutron (1n). This is still hydrogen, since elements are defined by the number of protons in the nucleus – the atomic number. But this is a slightly different, heavier version of hydrogen, called hydrogen-2. The two versions are isotopes of hydrogen. Add another neutron and you have hydrogen-3, with one proton and two neutrons (1p, 2n).

The strong nuclear force works with protons, too (but not electrons). If you force your other proton very close to your nucleus, the attractive strong nuclear force overcomes the repulsive force. The proton sticks after all, and your hydrogen-3 nucleus has become a nucleus of helium-4, with two protons and two neutrons (2p, 2n).

ELECTRON SHELLS

To the helium-4 nucleus you made, you will need two electrons if you want it to become a helium atom. Drop them in towards your new nucleus and they both occupy the fuzzy spherical orbital. The two electrons are both at the same energy level, n=1, and this spherical orbital a called an s-orbital – so this is the 1s orbital.

An orbital can hold up to two electrons, so when it comes to the third element, lithium, a new one is needed. This second orbital is another spherical s-orbital, about twice the size of the first, and it is at the next energy level, n=2. It is a 2s orbital. If you look at the periodic table, you will see that lithium is in the second row, or period. The rows of the periodic table correspond to the energy levels in which you find an atom's outermost electrons. The Period 2 elements, from lithium to neon, all have a full 1s orbital, and their outermost electrons at energy level n=2.

Electrons that share the same energy level around the nucleus of an atom are said to be in the same shell. The electrons of hydrogen and helium can fit within the first shell (energy level n=1). At the second energy level – in the second shell – there is more space for electrons, and a new type of orbital, the dumbbell-shaped p-orbital, makes its first appearance. There are three p-orbitals, giving space for up to six electrons. So the second shell contains a total of eight electrons – that is why there are eight elements altogether in the second row of the periodic table. Period 3 of the periodic table holds another eight elements. From shell 4 (Period 4) onwards five d-orbitals are available in addition to s- and p-orbitals. And from shell 6, seven f-orbitals are also available.

An atom is most stable when its outermost shell of electrons is full. Atoms with unfilled outer shells can easily lose, gain or share electrons, and so attain a full outer shell. This swapping or sharing of electrons leads to bonds between atoms. An atom that has lost or gained one or more electrons becomes a positively or negatively charged ion, and ions of opposite charge stick together, forming ionic bonds. Atoms that share electrons form covalent bonds, forming molecules.

Chemical reactions involve the breaking and/or forming of ionic and covalent bonds. This explains why the elements in Group 18, the far right of the periodic table, are unreactive – they have no need to lose, gain or share electrons, and so do not easily form bonds.

UNSTABLE NUCLEI

As we have been filling the electron shells, we should also have been adding protons to the nucleus, since the number of electrons in an atom is equal to the number of protons in the nucleus. So by now, the nuclei are much bigger than those of hydrogen or helium. Argon, with 18 electrons, must also have 18 protons in the nucleus. If a nucleus that big consisted only of protons, the protons' mutual repulsion would overpower the attraction of the strong nuclear forces. The nucleus would be extremely unstable and would fly apart in an instant. Neutrons provide the attractive strong nuclear force without adding the repulsive electrostatic force: they act like nuclear glue.

So, for example, the most common isotope of argon has 22 neutrons to help its 18 protons adhere. In other words, argon-40 (18p, 22n) is a stable isotope. There are other stable isotopes of argon, including argon-38 (18p, 20n) and argon-36 (18p, 18n). The average atomic mass (the standard atomic weight) of any sample of argon atoms is not a whole number; it is 39.948. No element has a standard atomic weight that is a whole number.

Generally, the largest nuclei are the most unstable – and elements with atomic numbers above 82 have no stable isotopes at all. There are several things that can happen to an unstable nucleus. The two most common are alpha decay and beta decay. In alpha decay, a large and unstable nucleus expels a clump of two protons and two neutrons, called an alpha particle. The atomic number reduces by two, because the nucleus loses two protons. So, for example, a nucleus of radium-226 (88p, 138n) ejects an alpha particle to become a nucleus of radon-222 (86p, 136n). Alpha decay results in a transmutation of one element

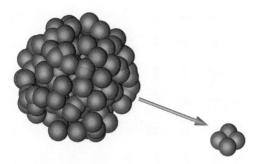

*Alpha decay. An unstable nucleus
loses an alpha particle (2p, 2n), reducing its
atomic number by 2.*

into another – in this case, radium becomes radon. The other main kind of decay is beta decay, in which a neutron spontaneously changes into a proton and an electron. The electron is expelled from the nucleus at high speed, as a beta particle. This time, the atomic number increases by one,

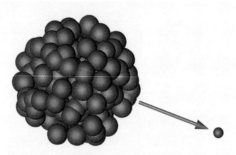

Beta decay. A neutron in an unstable nucleus spontaneously turns into a proton and an electron. The electron is released, as a beta particle.

since there is now an extra proton in the nucleus. Alpha and beta decay are random processes, but in a sample of millions or billions of atoms, the time for half of them to decay is always the same; this is called the half-life.

Nuclear instability, leading to radioactive decay,

is the reason there are no more than about 90 naturally occurring elements. Any heavier ones that were made billions of years ago have long since disintegrated to form lighter elements. Some, like uranium, have no stable isotopes, but do exist naturally, as some of their isotopes have half-lives of billions of years, so they still exist. Elements heavier than uranium, element 92, have only been made artificially, and most have only a fleeting existence – the transuranium elements.

Every element that exists naturally or has been created in laboratories – up to element number 118 – is featured in this book. More space is given to those elements that are particularly important or interesting. The book is mostly divided according to the vertical columns of the periodic table, called groups. Elements that are in the same group have very similar properties, because their outermost electron shell has the same electron configuration.

THE DATA FILES

Accompanying the profile of each element in this book is a data file giving four pieces of data:

ATOMIC NUMBER

This is simply the number of protons in the nucleus of each atom of that element. Atomic numbers range from 1 for hydrogen to 118 for oganesson, the heaviest element so far discovered.

ATOMIC WEIGHT

This is a shorthand for the more accurate term "standard atomic weight". Different isotopes (versions of the element's atoms with varying numbers of neutrons in the nucleus) have different weights. The standard atomic weight is an average of the atomic weights of all the isotopes of an element. In the case of the transuranium elements, whose isotopes are not well known, a whole number representing the atomic weight of the best-known isotope is given in parentheses.

MELTING POINT AND BOILING POINT

These are given in degrees Celsius (the unit many people still call "centigrade") and in Fahrenheit, and are as measured at average atmospheric pressure. Scientists normally use degrees Kelvin; the Kelvin scale starts at the coldest possible temperature (absolute zero), which is −273.15°C (−459.67°F). Since it is not familiar in everyday use, the Kelvin scale is not used in the data tables of this book.

CHAPTER

2

A
HISTORY

Ancient peoples were familiar with several of the substances that we now know as chemical elements. Some, such as gold, silver and sulfer, exist naturally; other, such as iron, copper and mercury, are easily extracted from minerals. But it was not until the end of the 18th century that scientists established the notion of what a chemical element actually is.

In many early civilizations, philosophers proposed that all matter is made up of four "elements" – earth, air, fire and water – in varying mixtures. This idea formed the basis for the mystical art of alchemy, whose main aim was the transformation of "base metals" such as lead into gold. Alchemy was as practical as it was mystical, and many of the basic techniques used by chemists to this day were developed by alchemists – as well as by apothecaries (pharmacists), glassmakers and metallurgists.

The flaws in the philosophical basis of alchemy were exposed by the scientific method, which became popular in Europe in the 17th century. Scientists gradually came to accept an idea that had been around for hundreds of years, but had never quite caught on: that matter is made of atoms. In the 1660s, Anglo-Irish scientist Robert Boyle used this idea to define a chemical element as a substance made of particles that are "primitive and simple, or perfectly unmingled". Towards the end of the 18th century, the concept of an element came into sharp focus thanks to the insight of French chemist Antoine Lavoisier, who proposed that an element should be defined simply as a substance that cannot be decomposed.

Lavoisier's insight into chemical elements was largely a result of his careful weighing of the reactants and products in chemical reactions, which showed that no mass is lost – the substances involved are simply rearranged. In 1808, English chemist John Dalton proposed that all the atoms of a particular element are identical, but different from those of other elements. The crucial, measurable difference was the mass of the atoms: hydrogen atoms are the lightest, sulfur are heavier and iron heavier still. This made sense of the fact that compounds are always composed of fixed ratios of substances by mass.

The rise of the scientific approach to chemistry helped scientists discover many chemical elements in the 19th century. Previously unknown metals were isolated from their "earths" (oxides), while the electric battery, invented in 1799, allowed chemists to discover elements using electrolysis ("electrical splitting"). The spectroscope (invented in the 1860s) allowed chemists to identify elements by studying the spectrum of light given out when substances are vaporized and heated; the presence of unfamiliar lines in spectra led to the discovery of several previously unknown elements. By this time, chemists began to realize that the growing list of elements seemed to fall into groups according to their properties. Furthermore, elements in the same group seemed to be spaced eight or 18 elements apart in a list of the elements by atomic weight. Russian chemist Dmitri Mendeleev made sense of this "periodicity" when he drew up the first periodic table, in 1869.

The discoveries of the electron, radioactivity and X-rays in the 1890s began a dramatic era in the development of atomic physics in the first half of the 20th century. The atomic nucleus was discovered in 1911, soon followed by its constituent parts the proton (1917) and the neutron (1932). Starting in the 1910s, quantum physics began to make sense of the arrangement and behaviour of electrons in atoms. The theories and experiments of nuclear physics enabled physicists to work out how elements are created, by the fusion of protons and neutrons inside stars and supernovas. Nuclear physics also led to the creation of elements heavier than uranium – the transuranium elements – most of which do not exist naturally.

HYDROGEN

Hydrogen is officially in Group 1 of the periodic table, since it has a single electron in its outermost shell. But that electron is its only one, so hydrogen has very different properties from the other Group 1 elements, and it is generally considered in a category of its own. Hydrogen is by far the most abundant element in the universe.

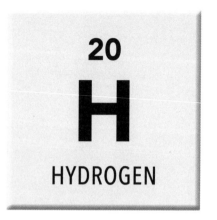

Atomic Number   1
Atomic Weight   1.008 g mol^{-1}
Melting Point   −259.14°C (−434.45°F)
Boiling Point   −252.87°C (−423.17°F)

HYDROGEN'S NAME MEANS
"WATER GENERATOR".

Hydrogen is a diatomic gas (two atoms per molecule, H_2) at room temperature; all the other Group 1 elements are solid metals. However, in the extreme pressures at the centre of gas giant planets such as Jupiter, hydrogen does behave like a metal. The vast clouds of dust and gas from which stars are born are mostly hydrogen, and this element is by far the most abundant in the universe.

English scientist Henry Cavendish is normally given credit for the discovery of hydrogen, after he produced and studied it in 1766. The gas Cavendish produced was explosive, and he suggested that it might be rich in a hypothetical substance that scientists of the day named "phlogiston". However, Cavendish couldn't explain why his "phlogisticated air" produced water when it burned. French chemist Antoine Lavoisier found the explanation in 1792, and named the element *hydrogène*, from the Greek for "water generator". Most hydrogen on Earth is combined with oxygen, in water molecules, H_2O.

Some water molecules naturally split, or dissociate, into two ions: hydrogen ions (H^+) and hydroxide ions (OH^-). Acidic solutions have greater concentrations of hydrogen ions than does pure water. The measure of acidity, known as pH, is actually a measure of the concentration of H^+ ions.

Fossil fuels, such as oil, coal and natural gas, consist mostly of hydrocarbons – molecules containing only carbon and hydrogen. When fossil fuels burn, oxygen atoms combine with the hydrocarbons, producing carbon dioxide (CO^2) and water (H_2O). Natural gas is the main source of hydrogen for industry. In a process called steam reforming, superheated steam separates the hydrogen from hydrocarbons such as methane (CH_4).

Almost two-thirds of all industrially produced hydrogen is used to make ammonia (NH_3), around 90 per cent of which in turn is used in the manufacture of fertilizers. Most of the rest of the hydrogen supply is used in processing crude oil, to help "crack" large hydrocarbon molecules into smaller molecules needed in commercial fuels and rid hydrocarbon molecules of unwanted sulfur atoms.

There are three isotopes of hydrogen. The most common, with a single proton as its nucleus, is referred to as protium. The only other stable isotope is deuterium (D), which has one proton and one neutron. The third isotope, tritium, has a proton and two neutrons.

Hydrogen isotopes are involved in experiments with nuclear fusion, based on the same reactions in the Sun and other stars. Fusion reactors here on Earth could provide a practically limitless supply of energy in the future. Fusion reactions involving deuterium and tritium are also the source of energy of hydrogen bombs. Inside an H-bomb, a conventional atomic bomb creates sufficiently high pressure and temperature for fusion to occur.

Even before nuclear fusion becomes viable, hydrogen may replace fossil fuels as a common energy "currency". Electricity from renewable sources can be used to separate it from water, through a process called electrolysis. The resulting hydrogen has a high energy density and can be stored and transported fairly easily. Most hydrogen-powered vehicles use hydrogen fuel cells, which rely upon a chemical reaction that is the reverse of electrolysis: hydrogen combines with oxygen. The reaction is the equivalent of burning hydrogen – the waste product is water – but is slower and more controlled, and produces electrical energy instead of heat.

CHAPTER

4

THE ALKALI METALS

In contrast with the uniqueness of hydrogen and the varied properties you find in some other groups of the periodic table, the alkali metals exhibit particularly strong family resemblances. All are solid, but soft, at room temperature; all are shiny metals that have to be kept under oil or in an inert atmosphere because they are very reactive.

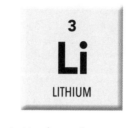

Atomic Number	3
Atomic Weight	6.94
Melting Point	180°C (357°F)
Boiling Point	1,345°C (2,448°F)

Lithium is the least dense solid element, and one of the most reactive metals. It was discovered in 1817, by Swedish chemist Johan Arfwedson, who was searching for compounds of the recently discovered element potassium. Arfwedson based the name of the element on the Greek word *lithos*, meaning "stone". Compounds of lithium are used in lithium ion rechargeable batteries for cameras, laptop computers and electric vehicles; in the glass and ceramics industry and in medicines for various psychological disorders.

11

Na

SODIUM

Atomic Number	11
Atomic Weight	22.99
Melting Point	98°C (208°F)
Boiling Point	883°C (1,621°F)

Humphry Davy discovered sodium in 1807, by passing electricity through molten caustic soda (sodium hydroxide). He derived the name from *sodanum*, the Roman name for plants whose ashes were used in glassmaking. The symbol for sodium, Na, comes from *natrium*, the Latin name for sodium carbonate. In ancient Egypt, sodium carbonate, called natron, was used as a drying agent in mummification. Important sodium compounds include table salt (sodium chloride) and sodium bicarbonate, used as a raising agent.

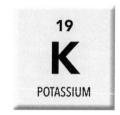

Atomic Number	19
Atomic Weight	39.10
Melting Point	63°C (146°F)
Boiling Point	760°C (1,398°F)

Humphry Davy discovered potassium in 1807, in the alkaline compound called potash (potassium hydroxide), which was and still is used to make soap. The element's symbol, K, is from the Latin word for alkali: *kalium*. Potassium hydroxide has many other uses in industry, and is used in alkaline batteries. Two other important potassium compounds are potassium chloride and potassium nitrate, used mostly in fertilizers. In humans and animals, potassium is key to nerve function. Fruits and vegetables contain plenty of potassium.

Atomic Number   37
Atomic Weight   85.47
Melting Point   39°C (103°F)
Boiling Point   688°C (1,270°F)

The German chemists Robert Bunsen and Gustav Kirchhoff discovered rubidium in 1861, using spectroscopy (*see page 34*). Pure rubidium was first extracted in 1928. Although rubidium is fairly abundant in the Earth's crust, it has few applications and is not essential in living organisms. The radioactive isotope rubidium-87 is used in medicine, enabling radiographers to pinpoint regions of low blood flow (ischaemia). In the future, rubidium may be used in ion drive engines for spacecraft exploring deep space.

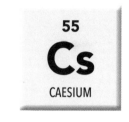

Atomic Number	55
Atomic Weight	132.91
Melting Point	28°C (83°F)
Boiling Point	671°C (1,240°F)

German chemists Robert Bunsen and Gustav Kirchhoff discovered caesium in 1861, using spectroscopy (*see page 34*). The name is from the Latin *caesius*, meaning "sky blue". Because of its use in atomic clocks, caesium was used in the definition of a second. In 1967, the International Committee on Weights and Measures (CIPM) decided that "The second is the duration of 9 192 631 770 periods of the radiation corresponding to the transition between the two hyperfine levels of the ground state of the caesium-133 atom."

Atomic Number	87
Atomic Weight	223
Melting Point	23°C (73°F), estimated
Boiling Point	680°C (1,250°F), estimated

Francium was discovered in 1939 by French physicist Marguerite Perey. She named it after her native country. Thirty-four isotopes are known, but even the most stable of them, francium-223, has a half-life of just 22 minutes. Nevertheless, francium does occur naturally, although probably not more than a few grams at any one time in the whole world: it is the product of the decay of other radioactive elements, most notably actinium (*see page 116*).

CHAPTER

5

THE
ALKALINE
EARTH
METALS

The alkaline earths are stronger, denser and less reactive than their Group 1 counterparts. In the Middle Ages, the term "earth" was applied to substances that do not decompose on heating, as is true of the oxides of calcium and magnesium. The "alkaline" part of the group's name relates to the fact that the oxides of these elements dissolve in water, producing alkaline solutions.

Atomic Number	4
Atomic Weight	9.01
Melting Point	1,287°C (2,349°F)
Boiling Point	2,469°C (4,476°F)

French chemist Nicolas-Louis Vauquelin discovered beryllium in a sample of beryl in 1798 – although no one prepared a sample of it for another 30 years. Unlike the other alkaline earth metals, beryllium does not form ions. As a result, beryllium compounds are all covalent, rather than ionic (*see page 22*). Beryllium is highly reflective of infrared light, and it can be worked very precisely to a polished finish, so it is used to make mirrors for orbiting infrared telescopes.

Atomic Number	12
Atomic Weight	24.31
Melting Point	650°C (1,201°F)
Boiling Point	1,091°C (1,994°F)

There are more than a million tonnes of magnesium per cubic kilometre of seawater. The human body contains about 25 grams of it. Green vegetables are good dietary sources of magnesium, since atoms of the element lie at the heart of the chlorophyll molecule. Scottish chemist Joseph Black first identified magnesium as an element, in 1755, while experimenting with magnesia alba (magnesium carbonate), an ore that was found in the Magnesia region of Greece. English chemist Humphry Davy first prepared the element, in 1808.

20

Ca

CALCIUM

Atomic Number	20
Atomic Weight	40.08
Melting Point	842°C (1,548°F)
Boiling Point	1,484°C (2,703°F)

In 1808, Humphry Davy discovered calcium, after separating it from a mixture of lime (calcium oxide) and mercury oxide. Davy named the element after the Latin word for lime, *calx*. Nearly all the calcium in the human body is tied up in calcium phosphate, the main constituent of bones and teeth. Dairy products are well known as good dietary sources of calcium. Vitamin D is required for the proper uptake of calcium from the gut; a deficiency of either calcium or vitamin D results in the disease rickets.

Atomic Number 38
Atomic Weight 87.62
Melting Point 777°C (1,431°F)
Boiling Point 1,382°C (2,520°F)

In 1790, Irish chemist Adair Crawford studied a rock that had been found in the Scottish town of Strontian. He surmised that it must contain a new element. In 1808, Humphry Davy isolated the element and named it strontium. Strontium compounds are used in fireworks and glow-in-the-dark toys. Strontium has no biological role, but the bones and teeth of everyone on Earth contain small amounts of it, because its chemical behaviour is very close to that of calcium.

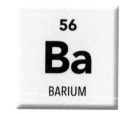

Atomic Number	56
Atomic Weight	137.30
Melting Point	727°C (1,341°F)
Boiling Point	1,897°C (3,447°F)

Barium, whose name derives from the Greek word *barys*, meaning "heavy", was first isolated by Humphry Davy, in 1808. More than 30 years earlier, German chemist Carl Wilhelm Scheele had realized that a previously unknown element was present in the mineral barite (barium sulfate). This compound has many uses, including in medicine. Barium is opaque to X-rays, and barium sulfate is used in a "barium meal" for patients undergoing X-ray imaging for digestive problems.

Atomic Number	88
Atomic Weight	226
Melting Point	700°C (1,292°F)
Boiling Point	1,737°C (3,159°F)

All isotopes of radium are highly radioactive. The element's name comes from the Latin word for ray, *radius*. Radium was discovered in 1898, by husband and wife team Marie and Pierre Curie. Marie and the French chemist André-Louis Debierne were the first to extract pure radium, in 1910. In the 1920s, before the dangers of radioactivity were understood, radium-containing items such as water soap – and even radium wool to keep babies warm – were marketed as health products.

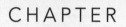

CHAPTER

6

THE
TRANSITION
METALS

The transition metals are elements that occupy the middle section (called the d-block) of the periodic table. They all have unfilled d-orbitals (*see page 21*). They are all shiny (lustrous) metals that are malleable (can be beaten thin) and ductile (can be stretched). Some of the most familiar metals are here – including copper, iron, gold and silver.

Atomic Number	21
Atomic Weight	44.96
Melting Point	1,541°C (2,806°F)
Boiling Point	2,836°C (5,136°F)

When Dmitri Mendeleev published the first periodic table (*see page 35*), he left several gaps. In 1879, scandium was assigned to one of those gaps, after Swedish chemist Lars Fredrik Nilson discovered it in a mineral found only in Scandinavia (hence the name). Of the few tonnes of scandium produced each year, most is used to make strong, lightweight alloys. It was not until 1937 that this elusive element was produced in its pure state.

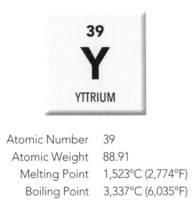

Atomic Number	39
Atomic Weight	88.91
Melting Point	1,523°C (2,774°F)
Boiling Point	3,337°C (6,035°F)

Finnish chemist Johan Gadolin discovered yttrium in a mineral from a quarry in the Swedish village of Ytterby. He identified the oxide of the new element in 1794. Yttrium was first isolated by German chemist Friedrich Wöhler in 1828. Yttrium aluminium garnet (YAG) lasers are used in a wide variety of settings, and yttrium oxide added to zirconium oxide produces a very stable and inert ceramic used in heat-resistant elements in jet engines and industrial abrasives and bearings.

Atomic Number	22
Atomic Weight	47.87
Melting Point	1,665°C (3,029°F)
Boiling Point	3,287°C (5,949°F)

Titanium was first detected in an unknown oxide in the 1790s. Its name derives from the Titans of Greek mythology. Titanium is reactive, but an oxide layer forms, making it very unreactive. Its inertness means it is ideal for medical implants, such as artificial hip joints. Strong titanium alloys are used in the aerospace industry. The most important titanium compound is titanium oxide. This bright white powder has many uses, including as a pigment in paints and cosmetics.

Atomic Number	40
Atomic Weight	91.22
Melting Point	1,854°C (3,369°F)
Boiling Point	4,406°C (7,963°F)

Zirconium was discovered in 1789, when German chemist Martin Klaproth produced an oxide from the mineral zircon. Swedish chemist Jöns Jacob Berzelius produced nearly pure zirconium in 1824. Alloys of zirconium are very hard and durable, and are used in situations where components can come under extreme heat and pressures, such as in nuclear reactor assemblies. Zirconium oxide is known as zirconia, One form, known as cubic zirconia, is used in jewellery as a cheaper alternative to diamonds.

Atomic Number	72
Atomic Weight	178.49
Melting Point	2,233°C (4,051°F)
Boiling Point	4,600°C (8,312°F)

Hafnium is another silver-grey metal. Its existence was predicted, as it filled a gap in the periodic table. It was discovered in Copenhagen in 1923 – Hafnia is the Latin name for Copenhagen – by Hungarian chemist George de Hevesy and Dutch physicist Dirk Coster. Hafnium is used in control rods in nuclear reactors, and in certain very hard alloys. One such alloy was used for the rocket nozzles on the Apollo Lunar Module.

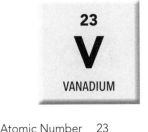

Atomic Number 23
Atomic Weight 50.94
Melting Point 1,910°C (3,470°F)
Boiling Point 3,407°C (6,165°F)

Spanish mineralogist Andrés Manuel del Río discovered vanadium in 1801. The name is derived from Vanadis, which means "lady of the Vanir tribe" and refers to the Norse goddess Freyja. Vanadium is a typical transition metal, forming colourful compounds and strong, useful alloys. Vanadium is an essential element for humans, but no one knows exactly why. It seems to be a catalyst for certain metabolic reactions. Only tiny amounts are needed – any excess is excreted. Vanadium is found in shellfish, mushrooms and liver.

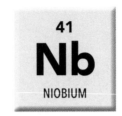

41

Nb

NIOBIUM

Atomic Number	41
Atomic Weight	92.91
Melting Point	2,477°C (4,491°F)
Boiling Point	4,744°C (8,571°F)

English chemist Charles Hatchett discovered this new element in 1801, in a sample of a mineral sent from Massachusetts. A year later, tantalum was discovered – and for a while, some chemists thought the two were one and the same. German chemist Heinrich Rose ended the confusion in 1844, and named this new element niobium, after Niobe, daughter of the mythological Greek king Tantalus. Niobium is used in heat-resistant "superalloys" suitable for special applications in the chemical and aerospace industries.

Atomic Number 73
Atomic Weight 180.95
Melting Point 3,020°C (5,468°F)
Boiling Point 5,450°C (9,842°F)

Tantalum gets its from the mythological character Tantalus, also the source of the word "tantalize"). Swedish chemist Anders Ekeberg chose the name after he discovered the tantalizingly unreactive element in 1802. This inertness, actually due to an oxide layer that forms, means it is ideal for making medical implants. The metal itself is a good conductor of electricity, but the oxide layer is an insulator – this is put to good use in tantalum capacitors, found on computer circuit boards.

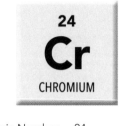

Atomic Number	24
Atomic Weight	52.00
Melting Point	1,860°C (3,380°F)
Boiling Point	2,672°C (4,840°F)

Nicolas-Louis Vauquelin derived chromium's name from the Greek word *chroma* ("colour"), after making coloured solutions from the mineral in which he discovered the element, in 1797. Electroplating chromium onto consumer items became popular in the 1950s. Chromium is the essential element in stainless steel, and is also used in the alloy nichrome, used as the heating element in hairdryers. The human body needs only tiny amounts; oysters, egg yolks and nuts are good sources.

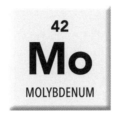

Atomic Number	42
Atomic Weight	95.94
Melting Point	2,620°C (4,750°F)
Boiling Point	4,640°C (8,380°F)

Molybdenum, another hard, grey transition metal, gives strength and heat resistance to a variety of alloys. Its name comes from molybdenite, the mineral in which it was discovered, in the 1770s. Swedish chemist Peter Hjelm was first to isolate the metal, in 1881. Molybdenite was named after the Greek word *molybdos*, meaning "lead". Humans require a small amount as part of their diet. Good sources of molybdenum include pulses such as lentils, and certain meats, including pork.

74

W

TUNGSTEN

Atomic Number	74
Atomic Weight	183.84
Melting Point	3,415°C (6,180°F)
Boiling Point	5,552°C (10,030°F)

Tungsten – from the Swedish for "heavy stone" – was the name of the mineral in which this metal was discovered, by Carl Wilhelm Scheele, in 1781. Tungsten has the highest melting point and boiling point of any metal, and is used as the filament of incandescent light bulbs. The symbol "W" relates to tungsten's alternative name, wolfram – from the German for "wolf spittle". Tungsten carbide is used to make hard machine tools, jewellery and the balls of ballpoint pens.

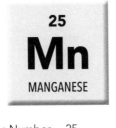

Atomic Number	25
Atomic Weight	54.94
Melting Point	1,245°C (2,273°F)
Boiling Point	1,962°C (3,563°F)

An adult human body contains around 12 milligrams of manganese, and around 5 milligrams are required each day; good sources include eggs, nuts, olive oil, green beans and oysters. In larger amounts, however, manganese is toxic. Like magnesium, manganese gets its name from Magnesia, in Greece, the origin of the mineral in which both elements were discovered, by several chemists in the 1770s. Used in steel production, manganese is present in most steel.

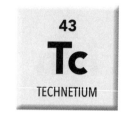

Atomic Number	43
Atomic Weight	(98)
Melting Point	2,170°C (3,940°F)
Boiling Point	4,265°C (7,710°F)

The name of this element is derived from the Greek word *tekhnitós*, which means "artificial"; technetium was the first element to be discovered only after being produced artificially. Very tiny amounts of technetium occur naturally, since every isotope of this element is unstable. It was discovered in 1937, by Italian mineralogist Carlo Perrier and Italian-American physicist Emilio Segrè, after it had been made artificially in an early particle accelerator.

Atomic Number	75
Atomic Weight	186.21
Melting Point	3,182°C (5,760°F)
Boiling Point	5,592°C (10,100°F)

In its pure state, rhenium is more dense than gold, with a very high melting point. It was discovered in 1925, by the German chemists Walter Noddack, Ida Tacke and Otto Berg, who named it after the River Rhine. Rhenium is rare, but is found in certain iron ores, and geologists use rhenium-osmium dating to date rocks more than a billion years old. Rhenium is also used in heat-resistant alloys and as a catalyst in the oil industry.

Atomic Number   26
Atomic Weight   55.84
Melting Point   1,535°C (2,795°F)
Boiling Point   2,750°C (4,980°F)

AN ADULT HUMAN BODY TYPICALLY CONTAINS
ABOUT 4 GRAMS OF IRON.

The Latin word for iron is *ferrum*, which is why the chemical symbol for iron is Fe. Pure iron is soft, silver-grey, malleable and ductile. Together with nickel and cobalt, iron is one of just three ferromagnetic elements: it can be magnetized, and is attracted to a magnet. Iron is by far the most magnetic of the elements, and even many iron ores can be magnetized.

Iron is a reactive metal, readily reacting with oxygen and water. One of the most familiar products is hydrated iron oxide, commonly known as rust. Mixing iron with various other elements can prevent it from rusting and change its properties in a host of other ways. Iron is the most common metal, and is used to make a huge range of alloys.

Before the dawn of the Iron Age, metal workers could only work with iron that had literally fallen from the sky – around one in 20 meteorites is made predominantly of iron. In the Middle East in the third millennium BCE, people began obtaining iron from its ores, by smelting. By around 1000 BCE, smelting was fairly common in several parts of the Middle East, Asia and Africa, spreading to Europe over the next few hundred years.

Early iron smelting took place in furnaces called bloomeries, in which charcoal burned at a high temperature as bellows pumped air into the fire. Carbon monoxide (CO) from the burning charcoal dragged oxygen away from the ore, leaving metallic iron. Molten iron fell to the bottom, forming an impure, porous mass called a bloom. Blooms had to be hammered – or wrought – to obtain fairly pure iron.

A different kind of furnace, the blast furnace, was invented in ancient China, and independently in Europe in the 15th century. In a blast furnace, limestone is added to the mix of iron ore and charcoal, which helps to remove impurities more effectively, while a strong blast of air enables a hotter burn. The liquid iron and the slag can be separated easily. Coke replaced charcoal in blast furnaces in the 18th century, helping enable the Industrial Revolution.

The molten iron from blast furnaces can be "cast" in moulds, but it is brittle. Wrought iron is malleable, but is too soft for many applications – and unlike cast iron, it rusts. The answer is to make steel – with more carbon than wrought iron and less than cast iron. Steel is hard, workable and less prone to corrosion.

Iron is a vital element in all living things, as many reactions rely upon it. Most iron in the human body is bound up in haemoglobin, the protein in red blood cells that is responsible for transporting oxygen in the bloodstream.

Iron accounts for around one-third of the mass of the planet Earth – the largest proportion of any element, followed closely by oxygen. In the Earth's crust, however, iron is only the fourth most abundant by mass, after oxygen, silicon and aluminium, because molten iron sank to the core early in the planet's history. As a result, iron makes up nearly 90 per cent of our planet's core (with nickel making up the rest) but only about 6 per cent of the crust. The inner core is solid, and it spins within the liquid outer core. This rotation is the source of the Earth's magnetic field, which stretches way out into space, creating a force field called the magnetosphere that protects us from the solar wind, a stream of ionized particles speeding out from the sun.

44

Ru

RUTHENIUM

Atomic Number	44
Atomic Weight	101.07
Melting Point	2,335°C (4,230°F)
Boiling Point	4,150°C (7,500°F)

Russian chemist Karl Klaus first identified and isolated ruthenium in a sample of platinum ore, in 1844. He based the name on "Ruthenia", a pseudo-Latin word for an ancient region of what is now western Russia and Eastern Europe. Like many transition metals, ruthenium is used in high-performance alloys and some ruthenium compounds make excellent catalysts. Ruthenium has no known biological role, and most ruthenium compounds are extremely toxic.

| 76 |
| **Os** |
| OSMIUM |

Atomic Number	76
Atomic Weight	190.23
Melting Point	3,030°C (5,485°F)
Boiling Point	5,020°C (9,070°F)

The name osmium comes from the Greek word *osme*, which means "smell"; osmium oxide has a pungent odour. English chemist Smithson Tennant discovered osmium in 1803, after dissolving a platinum mineral in strong acids. Osmium makes up just one part per billion of the Earth's crust by mass. It is mostly used in strong alloys. Osmium has the highest measured density of all elements: a block of osmium weighs twice as much as a block of lead of the same volume.

27

Co

COBALT

Atomic Number	27
Atomic Weight	58.93
Melting Point	1,495°C (2,723°F)
Boiling Point	2,900°C (5,250°F)

Cobalt, iron and nickel are the only elements that can be magnetized at room temperature, and cobalt is used in some permanent magnets. The name comes from Kobold, a mischievous sprite in German folklore. German miners used the name to refer to a mineral that was hard to smelt. The Swedish glassmaker Georg Brandt discovered the element in that mineral around 1735. Cobalt compounds are used as catalysts in the chemical industry, and the isotope cobalt-60 is used in radiotherapy.

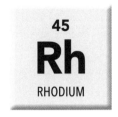

Atomic Number   45
Atomic Weight   102.91
Melting Point   1,965°C (3,570°F)
Boiling Point   3,697°C (6,686°F)

The name "rhodium" has the same origin as "rhododendron": the Greek word *rhodon*, meaning "rose". English chemist William Hyde Wollaston named the element, after he extracted the pure metal from a rose-coloured solution he had made from a platinum ore in 1803. The main use for rhodium since the 1970s has been in catalytic converters on vehicle exhausts, which help convert noxious carbon monoxide, hydrocarbons and oxides of nitrogen into less harmful substances.

77

Ir

IRIDIUM

Atomic Number	77
Atomic Weight	192.22
Melting Point	2,447°C (4,435°F)
Boiling Point	4,430°C (8,000°F)

Iridium was discovered in 1803 by the English chemist Smithson Tennant, at the same time as he discovered another of the platinum group, osmium. Because the new element produced many coloured compounds, Tennant named it after Iris, the Greek goddess associated with the rainbow. Iridium is the second most dense element, after osmium. It is used in a variety of alloys and catalysts – in particular where resistance to high temperature and wear is required.

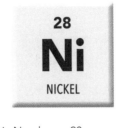

Atomic Number	28
Atomic Weight	58.69
Melting Point	1,453°C (2,647°F)
Boiling Point	2,730°C (4,945°F)

Swedish mineralogist Axel Cronstedt discovered nickel in a mineral called Kupfernickel, in 1751. It is used in a range of alloys with copper, called cupronickels, including the one of which the US dime is made. A nickel-copper-zinc alloy called nickel silver, popular with jewellery makers, is used to make zips. Along with iron and cobalt, nickel has some applications in magnetic materials, and compounds of nickel are used in nickel-cadmium and nickel metal hydride (NiMH) batteries.

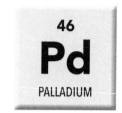

Atomic Number	46
Atomic Weight	106.42
Melting Point	1,552°C (2,825°F)
Boiling Point	2,967°C (5,372°F)

Palladium was discovered in 1802, by English chemist William Hyde Wollaston, in a solution he created by dissolving a platinum mineral in strong acids. He named it after a recently discovered asteroid, Pallas, from the Greek goddess of wisdom, Pallas Athena. Palladium is used in a range of alloys, and in catalytic converters for vehicle exhausts, and palladium catalysts are used to produce nitric acid. It is also used to make capacitors for smartphones and computers.

Atomic Number   78
Atomic Weight   195.08
Melting Point   1,769°C (3,216°F)
Boiling Point   3,827°C (6,920°F)

Platinum, one of the most expensive metals, exists naturally in a nearly pure state. People in what is now South America had used it in decorative items long before 1752, when Swedish assayist Henrik Scheffer identified it as an element and a precious metal. The name platinum comes from the Spanish word for silver, plata. Platinum is used in catalytic converters in vehicle exhaust systems. A platinum compound is used in the chemotherapy drug cisplatin, which has been used against a wide range of cancers since the late 1970s.

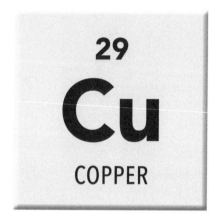

29

Cu

COPPER

Atomic Number	29
Atomic Weight	63.55
Melting Point	1,085°C (1,985°F)
Boiling Point	2,656°C (4,650°F)

.A TYPICAL CAR CONTAINS ABOUT
20 KILOGRAMS OF COPPER.

Copper occurs naturally in a fairly pure state and was known to ancient people; the oldest known artefacts made of copper are about 10,000 years old. Copper was first smelted from its ores around 7,000 years ago, and first combined with tin – making bronze – around 5,000 years ago. Across much of Europe and Asia, the Bronze Age lasted from around 3000 BCE until around 1000 BCE, when the alloy was largely superseded by iron and steel. Bronze is still used today, as is another ancient alloy, brass (copper and zinc). Bronze and brass were popular in the Roman Empire. The name copper is derived from Cuprum, the Latin name for Cyprus, where the Romans mined most of their copper.

There is a wide range of other copper alloys in use today, in addition to bronze and brass. The main uses of copper are as electrical wiring and water and gas pipes. Copper-rich alloys are being used increasingly on surfaces in hospitals and schools, because copper has a proven antimicrobial action. Copper is an essential element in humans; an adult requires around 1 milligram per day. Good dietary sources of copper include liver, egg yolks, cashew nuts and avocados.

Atomic Number	47
Atomic Weight	107.87
Melting Point	962°C (1,764°F)
Boiling Point	2,162°C (3,924°F)

SILVER CONDUCTS ELECTRICITY
BETTER THAN ANY OTHER MATERIAL AT
ROOM TEMPERATURE.

Silver is found in native (uncombined with other elements) form, so it was known to ancient metalworkers. Its chemical symbol, Ag, derives from its Latin name, *argentum*. When exposed to light, certain silver compounds blacken, as silver metal precipitates out. In 1801, German physicist Johann Ritter noticed that this still happened beyond the blue end of the spectrum, and thereby discovered ultraviolet radiation. In addition to its use in decorative items and coins, in the 20th century silver was very important in the development of photographic film.

With the rise of digital cameras, the demand for silver by the film and photographic industry has plummeted. But another application of silver is growing rapidly: solar cells. Silver is often used, suspended in a paste, to make fine connecting wires. Silver foil is also used to make the antennas in radio frequency identification tags (RFID), which are commonly attached to clothes, books and other consumer goods to prevent theft. Silver is rather toxic to micro-organisms, but not to humans, and silver compounds are used in some bandages. The metal is also used in some water purification filters, as dissolved silver ions that can kill bacteria.

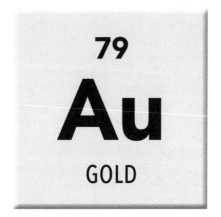

Atomic Number	79
Atomic Weight	196.97
Melting Point	1,064°C (1,948°F)
Boiling Point	2,840°C (5,144°F)

THE MOST EXPENSIVE DESSERT,
AT $25,000, WAS AN ICE CREAM SUNDAE
WITH GOLD LEAF.

The word "gold" comes from an old proto-Indo-European word, *ghel*, meaning "yellow". The chemical symbol Au is derived from the Latin for gold, *aurum*. Gold is found native (uncombined with other elements), most of it as tiny particles embedded in rocks. Rivers erode these rocks, and particles of gold wash downstream, in flakes and nuggets. When people pan for gold, they collect silt in a round-bottomed pan and agitate it, so that any gold falls to the bottom. Most gold, however – around 2,500 tonnes each year – is extracted by mining. In most countries, the purity of gold is measured in carats (symbol ct; in the USA it is kt or k, for karat). In theory, 24-carat gold is 100 per cent pure, although in practice a tolerance of 0.01 per cent is allowed.

As well as being used for decorative items, and as gold reserve currency, gold has many practical applications. It is normally the metal of choice for fine bonding wires that connect to semiconductor chips – its biggest industrial application. High-end audio connectors are gold or gold-plated, because gold does not tarnish and has very high electrical conductivity.

30

Zn

ZINC

Atomic Number	30
Atomic Weight	65.39
Melting Point	420°C (787°F)
Boiling Point	907°C (1,665°F)

Zinc was known to the ancient Romans, but only recognized as a metal in the 1520s – by Swiss physician Paracelsus, who used the German word *zinken* to describe the sharp projections of zinc crystals. Zinc is essential for living things – good sources include shellfish and cereals. Galvanizing – coating steel surfaces with zinc – protects from corrosion. Alloyed with copper, it makes brass. Zinc oxide is used in sunscreens, while zinc sulfide is used to make "glow-in-the-dark" pigments.

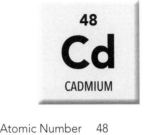

48

Cd

CADMIUM

Atomic Number	48
Atomic Weight	112.41
Melting Point	321°C (610°F)
Boiling Point	767°C (1,413°F)

In 1817, German chemist Friedrich Stromeyer discovered a silver-blue metal in a sample of calamine (zinc carbonate). He named it after the Latin word for the mineral calamine: *cadmia*. Cadmium-based paints were popular in the 19th century, until scientists realized cadmium is toxic. Since the 1980s, the main use for cadmium has been in nickel-cadmium (NiCad). Today, nickel metal hydride batteries are more popular, because they are cheaper and do not contain toxic cadmium.

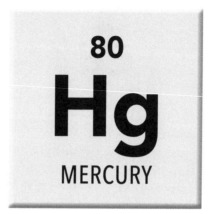

Atomic Number	80
Atomic Weight	200.59
Melting Point	−39°C (−38°F)
Boiling Point	357°C (675°F)

THE SYMBOL Hg COMES FROM
THE LATIN WORD *HYDRARGYRUM*, MEANING
"LIQUID SILVER".

Mercury is named after the messenger god of the same name in Ancient Roman mythology. It is the only metallic element that is liquid at room temperature; only one other element, non-metallic bromine, shares this property. Mercury does occur native (uncombined with other elements), but only rarely.

The mineral cinnabar (mostly mercury sulfide) was highly prized in many ancient civilizations, and was used to make the pigment vermilion. It was also important to alchemists – the mystical forerunners of modern chemists, to whom mercury represented the "principle" of spirit.

As alchemy began to give way to science, mercury played a vital role in many experiments and discoveries – in particular, the mercury barometer (1643) and the mercury-in-glass thermometer (1714). Liquid mercury is a poor electrical conductor, but mercury vapour conducts well, and this is why it is used in energy-saving fluorescent lamps. Many other applications of mercury have disappeared, a result of concerns over its toxicity. Symptoms of chronic mercury poisoning include loss of hair, kidney malfunction, nervousness, insomnia and dementia.

CHAPTER

THE
LANTHANOIDS

The elements of the lanthanoid series make up the first row of the f-block of the periodic table. In their metallic form, lanthanoids are relatively soft metals that react vigorously with oxygen if left open to the air. No lanthanoids have any known biological role.

The lanthanoids, together with scandium and yttrium, are often referred to as the rare earth metals.

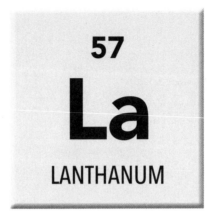

Atomic Number	57
Atomic Weight	138.91
Melting Point	920°C (1,688°F)
Boiling Point	3,464°C (6,267°F)

LANTHANUM IS USED IN NICKEL METAL
HYDRIDE BATTERIES IN HYBRID VEHICLES.

Lanthanum was discovered in 1839, by Swedish chemist Carl Mosander. Another Swedish chemist, Jöns Jacob Berzelius, suggested the name lanthana, from the Greek word *lanthano*, which means "lying hidden" – Mosander had found lanthanum as an impurity in the mineral cerite. In 1842, Mosander went on to discover that lanthanum was actually two elements; one kept the name lanthanum, and he called the other didymium, from the Greek *didymos*, meaning "twin". Didymium itself turned out to be three elements, known today as samarium, praseodymium and neodymium.

Lanthanum is the first element in the lanthanoid series – the top line of the f-block of the periodic table. The f-block (filled with elements whose outermost electrons are in f-orbitals) is normally shown separate from the main table, since otherwise the table becomes too unwieldily. Lanthanum is used in rechargeable nickel metal hydride batteries; an electric vehicle can contain several kilograms of this element. Lanthanum compounds are used in glassmaking, in the manufacture of high-quality lenses, and in phosphors.

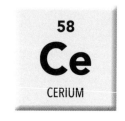

Atomic Number	58
Atomic Weight	140.12
Melting Point	799°C (1,470°F)
Boiling Point	3,442°C (6,227°F)

Cerium, the first of the lanthanoids to be discovered, was detected independently by the Swedish chemist Jöns Jacob Berzelius and German chemist Martin Klaproth, in 1803. It was named after the dwarf planet Ceres, which was discovered in 1801. Cerium is used as a catalyst in catalytic converters, and in self-cleaning ovens. It also makes up about 50 per cent of an alloy called misch metal, which is used as sparking flints in lighters.

Atomic Number   59
Atomic Weight   140.91
Melting Point   930°C (1,708°F)
Boiling Point   3,520°C (6,370°F)

In 1885, Austrian chemist Carl Auer von Welsbach investigated didymium (*see* lanthanum, *page 98*) and found that it was two elements. He called them praseodymium, from the Greek for "green-coloured twin", and neodymium ("new twin"). Compounds of these two are used as catalysts; both are also used in iron and magnesium alloys. "Didymium" now refers to a mix of praseodymium and neodymium, used for optical coating on lenses and filters. In safety goggles, it absorbs ultraviolet radiation.

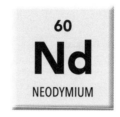

Atomic Number	60
Atomic Weight	144.24
Melting Point	1,020°C (1,868°F)
Boiling Point	3,074°C (5,565°F)

Neodymium was one of two elements that scientists had thought were one; its name derives from the Greek for "new twin". For more details about the discovery of neodymium, see praseodymium (*page 101*) and lanthanum (*page 98*). Neodymium is the crucial component of an alloy used to make strong permanent magnets. Because even small neodymium magnets are strong, they are the most widely used magnets in consumer electronics applications.

Atomic Number	61
Atomic Weight	(145)
Melting Point	1,042°C (1,907°F)
Boiling Point	3,000°C (5,430°F), estimated

With no stable isotopes, probably less than a kilogram of promethium exists naturally on Earth. Promethium was first produced in 1945, in a nuclear reactor at the Oak Ridge National Laboratory, USA. The name is from Prometheus, the Titan of Greek mythology who stole fire from the Zeus and brought it to humans. The first bulk sample of promethium metal was produced in 1963. Promethium has a few specialized applications, including use in some nuclear batteries for spacecraft.

62

Sm

SAMARIUM

Atomic Number	62
Atomic Weight	150.36
Melting Point	1,075°C (1,967°F)
Boiling Point	1,794°C (3,261°F)

French chemist Paul-Émile Lecoq de Boisbaudran discovered samarium in 1879, in samples of the mineral samarskite. Samarium is used as a catalyst in a process called SACRED (samarium-catalysed reductive dechlorination), which breaks down toxic chlorinated compounds. In an alloy with cobalt, samarium is used to make strong magnets, and it is one of the elements in a superconducting material – with zero electrical resistance – that could one day make power transmission and electric motors much more efficient.

Atomic Number   63
Atomic Weight   151.96
Melting Point   822°C (1,511°F)
Boiling Point   1,527°C (2,780°F)

Samarium is a soft grey metal, the most reactive of the lanthanoids. Paul-Émile Lecoq de Boisbaudran became aware of the new element in the spectrum of light from an impure sample of samarium, which he had discovered three years earlier. In 1901, French chemist Eugène-Anatole Demarçay extracted it, naming it europium, for Europe. The compound europium oxide glows bright red when illuminated by ultraviolet radiation, and is used in fluorescent lamps and liquid crystal display televisions.

64

Gd

GADOLINIUM

Atomic Number	64
Atomic Weight	157.25
Melting Point	1,313°C (2,395°F)
Boiling Point	3,272°C (5,921°F)

In 1880, Swiss chemist Jean de Marignac detected the spectrum of an unknown element in the light given off by a sample of the mineral gadolinite, named after Finnish chemist and geologist Johan Gadolin. French chemist Paul-Émile Lecoq de Boisbaudran named the element after he became the first to isolate it, in 1886. Gadolinium is very good at absorbing neutrons produced in nuclear reactions; as a result, it is used in shielding and control rods in nuclear reactors.

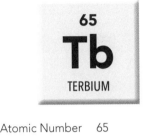

Atomic Number   65
Atomic Weight   158.93
Melting Point   1,356°C (2,473°F)
Boiling Point   3,227°C (5,840°F)

In 1843, Carl Mosander found two unknown elements, naming them terbium and erbium, after the mineral in which they were discovered, ytterbia – which in turn was named after the Swedish village of Ytterby. Due to confusion in the 1860s, the names became swapped. In 1886, Jean de Marignac discovered that terbium (Mosander's erbium) was actually two elements – gadolinium and modern-day terbium. The most important use of terbium today is in yellow-green phosphors used in fluorescent lamps.

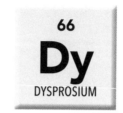

Atomic Number	66
Atomic Weight	162.50
Melting Point	1,412°C (2,573°F)
Boiling Point	2,567°C (4,653°F)

In 1886, French chemist Paul-Émile Lecoq de Boisbaudran discovered the oxide of an unknown element in a sample of the oxide of another lanthanoid, holmium (*see page 109*). He named the element dysprosium, after the Greek word *dysprositos*, meaning "hard to obtain"; after many attempts he managed to produce a small sample of it. Dysprosium has applications in lasers and lighting, and it is increasingly being used together with neodymium in strong magnetic alloys, particularly in motors for electric cars.

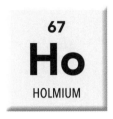

Atomic Number	67
Atomic Weight	164.93
Melting Point	1,474°C (2,685°F)
Boiling Point	2,700°C (4,885°F)

Holmium is a malleable silver-grey metal. In 1879, Swedish chemist Per Teodor Cleve discovered the oxides of three elements – thulium, erbium (*see pages 110-111*) and holmium – in what had been considered just one: erbium. The name holmium is from the Latin for Stockholm, Holmia. Today, holmium's primary use is in solid-state medical lasers. Holmium oxide is used to give a yellow-red colour to glass and to cubic zirconia stones in jewellery.

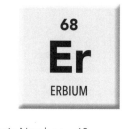

68

Er

ERBIUM

Atomic Number	68
Atomic Weight	167.26
Melting Point	1,529°C (2,784°F)
Boiling Point	2,868°C (5,194°F)

The name erbium was originally used for the element terbium (*see page 107*). In 1879, Per Teodor Cleve found that what chemists had thought was pure erbium oxide also contained the oxides of two other elements (thulium, *see page 111*, and holmium, *see page 109*). Erbium is used in lasers, and its oxide is added in tiny amounts to cubic zirconia, to give a pink colour. Erbium is also used as a dopant in fibre-optic cables, where it reduces the loss of signal strength.

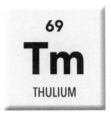

69

Tm

THULIUM

Atomic Number	69
Atomic Weight	168.93
Melting Point	1,545°C (2,813°F)
Boiling Point	1,950°C (3,542°F)

Thulium was one of the elements Per Teodor Cleve discovered in a sample of erbium oxide in 1879 (*see* erbium, *page 110*). Cleve named the new element after the ancient word for a mysterious, northerly part of Scandinavia, Thule. Like many of the lanthanoids, thulium is used in the crystals in solid-state lasers – and, like samarium, it may one day be used in superconducting materials. The radioactive isotope thulium-170 is used as an X-ray source for cancer treatment.

Atomic Number	70
Atomic Weight	173.04
Melting Point	824°C (1,515°F)
Boiling Point	1,196°C (2,185°F)

Ytterbium is one of four elements named after the village of Ytterby. Swiss chemist Jean de Marignac discovered the element in 1878, while studying one of the others, terbium. Ytterbium turned out to be two elements; one is now called scandium. In 1907, French chemist Georges Urbain showed that even this newly purified ytterbium was a mixture, including a new element now known as lutetium (*see page 113*). Today, ytterbium is used in some stainless steels, in solar cells and lasers.

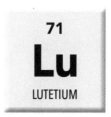

Atomic Number	71
Atomic Weight	174.97
Melting Point	1,663°C (3,025°F)
Boiling Point	3,400°C (6,150°F)

Element 71 was discovered in 1907, by French chemist Georges Urbain, in a sample of what was thought to be pure ytterbium (*see page 112*). Austrian mineralogist Carl Auer von Welsbach and American chemist Charles James independently discovered the element in the same year. Urbain proposed the name lutecium, with a "c", from the Latin name for Paris, *Lutetia*. The spelling was changed in 1949. Lutetium is used as a catalyst, and in sensors in medical scanners.

THE
ACTINOIDS

The elements of the actinoid series make up the second row of the f-block of the periodic table (*see also* the lanthanoid series, *pages 96–113*). They are all radioactive, and most of them are too unstable to exist naturally. Here we feature only those elements with atomic numbers less than 92 (uranium): the others are to be found in the transuranium elements section (*see pages 176–190*).

89

Ac

ACTINIUM

Atomic Number	89
Atomic Weight	(227)
Melting Point	1,050°C (1,922°F)
Boiling Point	3,200°C (5,790°F)

Actinium was discovered in 1899, by French chemist André-Louis Debierne, in the uranium ore from which Pierre and Marie Curie had discovered radium a year earlier. Debierne gave the new element the name "actinium" from the Greek word for ray, *aktinos*; the Latin equivalent, *radius*, had been used to name radium. Actinium has only a few, specialized, applications because of its rarity and radioactivity.

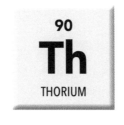

Atomic Number 90
Atomic Weight 232.04
Melting Point 1,750°C (3,190°F)
Boiling Point 4,790°C (8,650°F)

Thorium and uranium are the only actinoids whose atoms have survived in significant numbers since they were created in supernovas billions of years ago. Thorium was discovered by Swedish chemist Jöns Jacob Berzelius, in 1829. Berzelius chose the name "thorium" – derived from the Norse god of war, Thor. Thorium oxide has a very high melting point, and is used in producing heat-resistant ceramics and welding electrodes. Thorium itself is sometimes used in high-performance alloys with magnesium.

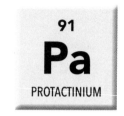

Atomic Number	91
Atomic Weight	231.04
Melting Point	1,572°C (2,862°F)
Boiling Point	4,000°C (7,230°F), approximate

Tiny amounts of protactinium exist naturally, the result of the decay of uranium nuclei. Protactinium was discovered in 1913 by Polish-American physicist Kazimierz Fajans and German physicist Oswald Göhring, who chose the name "brevium" – *brevis* is Latin for "short-lived". The name was changed to "protoactinium", because the element transmutes into actinium when it decays, and was shortened to its present form in 1949. Protactinium only has a few, specialized, applications because of its rarity and radioactivity.

Atomic Number	92
Atomic Weight	238.03
Melting Point	1,134°C (2,037°F)
Boiling Point	4,130°C (7,470°F)

By far the most important element in the actinoid series is uranium. One isotope of uranium, uranium-235, is the main source of fuel (as uranium oxide) in nuclear reactors. People had been using the uranium ore pitchblende as an additive in glassmaking for hundreds of years before uranium was recognized as an element. In 1789, German chemist Martin Klaproth realized that pitchblende contained an unknown element, which he named after the planet Uranus (discovered in 1781).

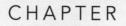

THE BORON GROUP

The atoms of all the elements in Group 13 have three electrons in their (incomplete) outer shell, so you might expect these elements to have very similar properties. But boron is a non-metal, forming only covalent bonds; aluminium and gallium are metallic, but with some non-metal characteristics; indium and thallium are true metals, and form only ionic bonds. This group also contains the transuranium element nihonium.

Atomic Number	5
Atomic Weight	10.81
Melting Point	2,079°C (3,774°F)
Boiling Point	4,000°C (7,232°F)

Boron is much rarer than most other light elements, because it was not made inside stars. The name "boron" derives from *buraq*, the ancient Arabic name for the mineral borax, which is used in detergents and cosmetics. Boron itself was first isolated from borax by several chemists in 1808. Heat-resistant glass is normally borosilicate glass, which is made with boron oxide. Boron is used to "dope" pure silicon in the manufacture of computer chips.

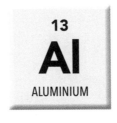

13

Al

ALUMINIUM

Atomic Number	13
Atomic Weight	26.98
Melting Point	660°C (1,220°F)
Boiling Point	2,519°C (4,566°F)

Aluminium gets its name from the mineral alum, from which it was first prepared, by several chemists, in 1808. The metal is often cast to make parts for buildings and vehicles, but can also be rolled into foil, drawn into wires and made into a powder. Aluminium powder is used in fireworks and in the manufacture of glass mirrors. Many aluminium compounds are also used in large quantities, including aluminium chlorohydrate, which is the most common active ingredient in antiperspirants.

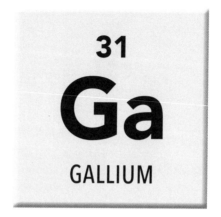

Atomic Number	31
Atomic Weight	69.72
Melting Point	30°C (86°F)
Boiling Point	2,205°C (4,000°F)

A SAMPLE OF SOLID GALLIUM METAL WILL
MELT ON THE PALM OF YOUR HAND.

Gallium was one of the elements whose existence Dmitri Mendeleev predicted in 1871, to fill the gaps in his periodic table. It was only four years before gallium was discovered, by French chemist Paul-Émile Lecoq de Boisbaudran. The name "gallium" is derived from the Latin word *Gallia*, which refers to the ancient land of Gaul, the precursor to modern-day France. Pure gallium is a shiny metal with a low melting point – just 30°C. An alloy called Galinstan®, containing around two-thirds gallium (along with indium and tin), melts at –19°C, and is increasingly used as a substitute for toxic mercury in thermometers.

Most of the demand for gallium is from the semiconductor industry, either for doping silicon to produce transistors, or for producing the compounds gallium arsenide and gallium nitride. The light-emitting crystals inside LEDs (light-emitting diodes) and laser diodes contain mixtures of several gallium compounds. Gallium arsenide is also used to make efficient solar cells, although the expense restricts their use to high-end applications such as providing power for satellites and space probes.

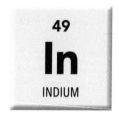

Atomic Number	49
Atomic Weight	114.82
Melting Point	157°C (314°F)
Boiling Point	2,075°C (3,770°F)

Indium was discovered in 1863, by German chemists Ferdinand Reich and Hieronymus Theodor Richter, using spectroscopy (*see page 34*). The name derives from the dominant colour of its spectrum, indigo. A few hundred tonnes of indium metal are produced each year, mostly as a by-product of the extraction of zinc. A mixture of indium and tin oxides is used to make transparent electrodes for flat-panel electronic displays. Another compound, indium nitride, is used along with gallium nitride in LEDs.

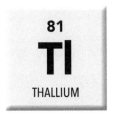

Atomic Number	81
Atomic Weight	204.38
Melting Point	303°C (577°F)
Boiling Point	1,473°C (2,683°F)

Like indium, thallium is named after the dominant colour in its spectrum, in this case green. The name is derived from the Greek word *thallos*, which means "green shoot". Thallium was discovered in 1861, independently, by the English physicist William Crookes and the French chemist Claude-Auguste Lamy. The element has a few applications, mostly in electronics. Thallium is highly toxic. The unpleasant symptoms of thallium poisoning include vomiting, hallucinations and loss of hair – and ultimately untimely death.

THE
CARBON
GROUP

In their elemental state, the members of the carbon group vary considerably: carbon is a black (or transparent, as diamond) non-metal; silicon and germanium are semiconducting metalloids; tin and lead are lustrous silvery metals. They form a wide variety of different compounds, and have an enormous range of applications. This group also includes the transuranium element flerovium.

6

C

CARBON

Atomic Number	6
Atomic Weight	12.01
Melting/Boiling Point	Sublimes at around 3,600°C (6,500°F)

MORE COMPOUNDS OF CARBON ARE KNOWN
THAN FOR ANY OTHER ELEMENT.

Pure carbon has several distinct forms, or allotropes. The main allotropes of carbon are diamond, graphite, graphene, carbon nanotubes and a range of substances called fullerenes. Diamonds are formed under high pressure and at high temperature, about 150 kilometres deep in the upper mantle, and brought to the surface by eruptions of magma. Graphite is made of layered flat planes of hexagonal rings of carbon atoms. Graphite is a conductor, and is used as the contacts in some electric motors. Graphite's planes are held together very loosely, and they can slip past each other and separate. This makes graphite a good lubricant – and is also why it is used in pencils (mixed with baked clay). Graphite can be made into thin fibres that, mixed with polymers, make tough carbon fibre-reinforced plastics.

Fullerenes are molecules consisting of carbon atoms joined in hexagonal and pentagonal rings. The first fullerene to be discovered, in 1985, was buckminsterfullerene, a spherical molecule containing 60 carbon atoms. Graphene is the equivalent of a single flat layer of graphite: carbon atoms arranged in hexagonal rings across a vast flat plane. Carbon nanotubes are like sheets of graphene rolled into cylinders. Together with graphene, they are set to play an important role in the future of electronics and materials science.

When combining with other elements, carbon is even more versatile, due to its ability to form single, double and triple bonds, and rings, and the ease with which it joins with other elements – particularly hydrogen, oxygen and nitrogen. Molecules with just carbon and hydrogen, called hydrocarbons, include methane and propane; candle wax is made of long-chain hydrocarbons.

Carbon compounds are intimately involved in all the processes of life, from photosynthesis and respiration through nutrition and repair to growth and reproduction. Because of this, carbon chemistry is called organic chemistry, although the importance of carbon compounds extends far beyond the realm of living things. Organic compounds form the basis of the petrochemical industry, and so include plastics and many synthetic dyes, adhesives and solvents.

In photosynthesis, plants use light energy to build the simple sugar glucose from carbon dioxide and water. The result is a store of chemical energy that can be utilized via respiration, by both plants and the animals that consume them. As well as using glucose for energy, plants combine glucose molecules to form larger organic molecules, such as cellulose. Other structural molecules, such as proteins, are also carbon-based. Chemicals involved in maintaining and repairing living organisms including enzymes, hormones, vitamins and DNA are all organic molecules.

The constant interchange of carbon through living systems, as well as non-living systems, is called the carbon cycle. The main stage of this cycle involves carbon dioxide being absorbed from the atmosphere during photosynthesis and then being released back into the atmosphere by respiration. When an organism dies, it typically decomposes – a process that results in its carbon content being released, as carbon dioxide or methane.

In some circumstances, an organism does not decompose. Its carbon content may slowly form mixtures of hydrocarbons: fossil fuels such as oil and coal. Since humans began burning fossil fuels, the carbon cycle has been out of balance, and atmospheric carbon dioxide continues to rise steeply.

Atomic Number	14
Atomic Weight	28.09
Melting Point	1,410°C (2,570°F)
Boiling Point	2,355°C (4,270°F)

SILICON IS THE SECOND MOST
ABUNDANT ELEMENT IN THE EARTH'S CRUST,
AFTER OXYGEN.

Silicon is best known for its applications in electronics. Most integrated circuits ("silicon chips") are made from wafers of ultra-pure silicon, doped with other elements. Ultra-pure silicon is also the basis of around 90 per cent of solar cells. Silicon most commonly exists in minerals called silicates. Silicate rocks are used to make bricks, ceramics and cement. Silicon is the basis of polymer compounds called silicones, which are used in rubbery, heat-resistant cookware, and in a liquid or gel form in household sealants and breast implants. The most useful compound of silicon is silica (silicon dioxide), which is the main component of quartz. Silica, mostly from (quartz) sand, is the main ingredient in glassmaking.

Our ancestors used one form of quartz, known as flint, to make axes and arrowheads; the name "silicon" is derived from the Latin word *silex*, meaning "flint". Silicon was first produced in its elemental form – and recognized as an element – in 1824, by Swedish chemist Jöns Jacob Berzelius.

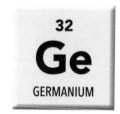

Atomic Number	32
Atomic Weight	72.63
Melting Point	938°C (1,721°F)
Boiling Point	2,834°C (5,133°F)

Like silicon, germanium is used in electronics, though much less so. Germanium's existence was predicted by Dmitri Mendeleev, as an element that would fill the gap below silicon in his periodic table. German chemist Clemens Winkler discovered the missing element in 1886. He derived the element's name from the Latin name for his home country, Germania. The most important compound of germanium is germanium dioxide, which has various applications, including in optical fibres and as a catalyst.

50

Sn

TIN

Atomic Number	50
Atomic Weight	118.71
Melting Point	232°C (450°F)
Boiling Point	2,590°C (4,695°F)

Tin was known to ancient metalworkers. The Latin word for tin, *stannum*, is the origin of this element's chemical symbol, Sn. The alloy of copper and tin known as bronze has been produced for at least 5,000 years. Pewter is another well-known tin alloy, typically consisting of 90 per cent tin. In its elemental form, tin is applied as a thin protective layer on other metals; most food cans are made from tin-plated steel, for example.

Atomic Number	82
Atomic Weight	207.20
Melting Point	327°C (621°F)
Boiling Point	1,750°C (3,182°F)

AROUND HALF OF ALL LEAD PRODUCED
IS USED TO MAKE LEAD-ACID CAR BATTERIES.

Until the second half of the 20th century, lead was the material of choice for plumbing pipes. The word "plumbing" comes from the Latin word for lead, *plumbum*, as does the element's symbol, Pb. The word "lead" simply comes from an Old English word for the metal. In its elemental form, lead is a soft, silvery-blue metal with a high lustre. Like many metals, lead quickly tarnishes when exposed to the air, giving it a dull grey appearance. The layer of tarnish protects it against corrosion. This, together with its malleability, has made lead suitable as a roofing material for hundreds of years, as it still is today. It is well known that lead is toxic.

The greatest danger comes from chronic (long-term) exposure to lead; like many heavy metals, it accumulates in tissues of the body, disrupting the nervous system. As a result of concerns over lead in the environment, many applications of lead have declined or been banned, including lead-based paints and leaded petrol.

THE
NITROGEN
GROUP

As is true for the elements of Group 14, the elements of Group 15 vary considerably in their properties – in this case, from the non-metals nitrogen and phosphorus, through the metalloids arsenic and antimony, to the "poor metal" bismuth. This group also includes the transuranium element moscovium.

Atomic Number   7
Atomic Weight   14.01
Melting Point   −210°C (−346°F)
Boiling Point   −196°C (−320°F)

NITROGEN GAS MAKES UP 78 PER CENT
OF THE ATMOSPHERE.

English scientist Henry Cavendish was the first to isolate the gas, French chemist Antoine Lavoisier the first to realize that it must be an element. Scientists soon found that the new gas could make nitric acid, which in turn could make nitre (potassium nitrate), an ingredient of gunpowder. The name *nitrogène* – meaning "nitre generator" – was suggested by the French chemist Jean-Antoine Chaptal in 1790.

Nitrogen is is present in many of the molecules of life, including all proteins and DNA. However, only a few organisms – bacteria called diazotrophs – are able to "fix" nitrogen. These organisms react nitrogen with hydrogen, to make ammonia. Together, diazotrophs fix an estimated 200 million tonnes of nitrogen worldwide each year.

In 1908, German chemist Fritz Haber developed a process for producing ammonia using atmospheric nitrogen, and another German chemist, Karl Bosch, scaled up the process. Today, the Haber–Bosch process fixes around 100 million tonnes of nitrogen each year – most of it used to produce artificial fertilizers. Ammonia produced by the process is also used to make plastics, pharmaceuticals and explosives.

15

P

PHOSPHORUS

Atomic Number	15
Atomic Weight	30.97
Melting/Boiling Point	Different for each allotrope

The non-metal phosphorus was the first element whose discovery is documented. German alchemist Hennig Brand found it, while trying to extract gold from urine. He named the new substance after the Greek word *phosphoros*, which means "torch-bearer". The element glows when it reacts with oxygen. In nature, phosphorus is nearly always bound to four oxygen atoms, making a phosphate ion, which is found in a wide range of phosphate minerals, and in all living things.

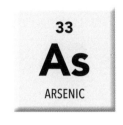

Atomic Number   33
Atomic Weight   74.92
Melting/Boiling Point   Sublimes at around
                          615°C (1,137°F)

Arsenic has two main allotropes: grey arsenic is
dense and lustrous like metals; yellow arsenic is
a crumbly non-metallic powder. Arsenic minerals
were used as pigments from ancient times. The
element's name comes from the Persian word for
one of them: zarnik. Today, most arsenic is used
in alloys with copper and lead and, with gallium,
in electronics. Most arsenic compounds are
highly toxic, and were used as poisons until the
beginning of the 20th century.

Atomic Number	51
Atomic Weight	121.76
Melting Point	631°C (1,168°F)
Boiling Point	1,587°C (2,889°F)

Antimony is similar to arsenic: it has metallic and non-metallic allotropes and it is highly toxic. Antimony and its compounds have been known for thousands of years. In ancient Egypt, a black mineral containing antimony sulfide was used as mascara. In Ancient Rome, that mineral was known as antimonium, while an alternative Latin term, *stibium*, gives the symbol. Today, the biggest use of antimony is alloys, most often with lead, typically used for the electrodes of car batteries.

Atomic Number	83
Atomic Weight	208.98
Melting Point	271°C (521°F)
Boiling Point	1,564°C (2,847°F)

Pure bismuth is a dense, shiny silver-white substance, like a metal – although it is brittle and it conducts electricity poorly. It is sometimes referred to as a "poor metal". The element's name may have come from the Greek word *psimythion*, meaning "white lead". Bismuth has few major applications. It is used in various alloys, and its compounds are used in some medicines and cosmetics. It is slightly toxic, but nowhere near as much as arsenic and antimony.

THE
OXYGEN
GROUP

The most important and abundant elements of Group 16 of the periodic table are the non-metals oxygen and sulfur. Selenium is also a non-metal, while tellurium is a metalloid. Polonium is sometimes considered a true metal, sometimes a metalloid. It is a radioactive element, with no stable isotopes. This group also includes the transuranium element livermorium.

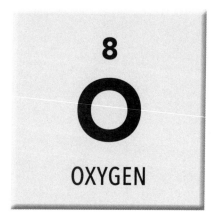

Atomic Number   8
Atomic Weight   16.00
Melting Point   −219°C (−362°F)
Boiling Point   −183°C (−297°F)

OXYGEN ACCOUNTS FOR TWO-THIRDS OF THE
MASS OF THE HUMAN BODY.

Oxygen is the third most abundant element in the universe, after hydrogen and helium. It is the second most abundant element on Earth as a whole, after iron – and by far the most abundant element in the Earth's crust, where it occurs mostly in oxide, silicate, carbonate and sulfate minerals. Oxygen also accounts for just under 90 per cent of the mass of pure water, and dissolved oxygen increases still further the amount of the element present in water. In addition, oxygen is the second most abundant element in the atmosphere, after nitrogen.

Oxygen is one of the most reactive of all the elements; that is why so much of it exists combined with other elements – in water, rocks and carbon dioxide, for example. The only reason the atmosphere and oceans contain large amounts of elemental oxygen is that certain living things constantly replenish it, as a waste product of photosynthesis. Oxygen atoms are also present in many other molecules involved in living processes, including all proteins, fats and DNA.

Elemental oxygen exists in the atmosphere in another form (allotrope), ozone, whose molecules have three oxygen atoms each. A tiny fraction of the total atmosphere, most ozone is found between 20 and 30 kilometres above ground – the ozone layer – where it intercepts most of the ultraviolet radiation that would be harmful to living things on Earth.

Oxygen is an essential part of combustion (burning). Most of the things we commonly burn – including wood and fossil fuels – are rich in carbon and hydrogen. When they burn, oxygen from the air combines with hydrogen to produce water, and with carbon to produce carbon dioxide. Inside the cells of most living things, a series of controlled combustion reactions occurs, called aerobic respiration, by which the organisms gain the energy they need to survive.

It was oxygen's role in combustion and respiration that led to the element's discovery, in the 1770s. English chemist Joseph Priestley was the first to publish anything about the discovery, in 1775. It was French chemist Antoine Lavoisier who realized that oxygen is a chemical element, and who gave the element its name. Lavoisier mistakenly believed that all acids contained the new element, so he came up with the name *oxygène*, meaning "acid former".

Being so abundant and reactive, oxygen is involved in some way in most processes, both natural and industrial – albeit mostly in compounds. So, for example, many metal ores are oxides of the metal, in which case extraction of the pure metal involves reactions that remove the oxygen.

Millions of tonnes of oxygen are produced industrially each year, from the distillation of liquefied air. More than half of it is used in steelmaking, and much of the rest in the chemical industry.

Some space-bound rockets carry liquid oxygen, held separate from the fuel. Some rockets derive the oxygen they need from a compound called an oxidizer, which releases large quantities of the gas when it heats up. The oxidizer is normally mixed in with the propellant, but in some cases, the propellant and the oxidizer are chemically combined. Explosives contain an oxidizer, too; in traditional gunpowder, for example, the oxidizer is saltpetre, or nitre (potassium nitrate).

Although oxygen gas appears colourless, it is actually a very pale blue at high pressures (although this is not why the sky is blue) and it is weakly magnetic.

Atomic Number	16
Atomic Weight	32.07
Melting Point	115°C (239°F)
Boiling Point	445°C (832°F)

Sulfur is found native (in its elemental state), in volcanic regions. Its name is simply the Latin word for it. Sulfur is an essential element for living things; in particular, it is present in many proteins. Sulfur dioxide is the precursor to one of the most important compounds in the modern chemical industry: sulfuric acid. Many sulfur-containing compounds are smelly. The smell of onions and garlic, and the unpleasant odour of flatulence, are all due to sulfur compounds.

Atomic Number	34
Atomic Weight	78.96
Melting Point	221°C (430°F)
Boiling Point	685°C (1,265°F)

Selenium is normally found in ores containing sulfur, including pyrite. It was in a sample of pyrite that selenium was discovered, in 1817, by Swedish chemists Jöns Jacob Berzelius and Johann Gahn. Berzelius named it after the Greek goddess of the moon, Selene. Selenium's main applications are in glassmaking and in the extraction of manganese. Selenium is an essential element in animals and some plants. Foods rich in selenium include Brazil nuts, tuna and sunflower seeds.

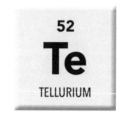

Atomic Number	52
Atomic Weight	127.60
Melting Point	450°C (841°F)
Boiling Point	988°C (1,810°F)

The main use of tellurium is in alloys with copper and lead, and the element is sometimes added to steel to make it more workable. Tellurium and some of its compounds are extremely toxic. Tellurium was discovered in 1782 by the Austrian mineralogist Franz-Joseph Müller von Reichenstein, in a sample of gold ore. In 1798, German chemist Martin Klaproth isolated pure tellurium from the ore, and named the new element after the Latin word *tellus*, meaning "Earth".

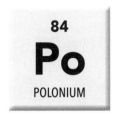

Atomic Number	84
Atomic Weight	(210)
Melting Point	254°C (489°F)
Boiling Point	962°C (1,764°F)

Polonium was discovered in 1898 by the Polish physicist Marie Curie and her husband, Pierre, while studying a sample of uranium ore. They named the element after Marie's country of birth. All isotopes of polonium are unstable. The longest-lived naturally occurring isotope, polonium-210, has a half-life of just 138 days. However, new polonium atoms are created constantly inside uranium ores, as part of a decay chain that begins with the disintegration of uranium atoms.

CHAPTER

13

THE
HALOGENS

All the Group 17 elements are reactive non-metals that form singly charged negative ions and similar compounds. However, there are differences when these elements are pure: fluorine and chlorine are gases at room temperature, bromine is liquid and iodine is solid. Astatine also stands apart – it is highly radioactive. This group includes the transuranium element tennessine.

9

F

FLUORINE

Atomic Number	9
Atomic Weight	19.00
Melting Point	−220°C (−363°F)
Boiling Point	−188°C (−307°F)

FLUORIDE COMPOUNDS ADDED TO
TOOTHPASTES REMINERALIZE TOOTH ENAMEL

At room temperature, fluorine is a pale yellow gas made of molecules of two fluorine atoms bonded covalently. All other bonds fluorine makes are ionic (*see page 22*), because fluorine atoms easily gain an electron, becoming negative fluoride ions. The most important fluoride-containing mineral is fluorite, also known as fluorspar. Since the 16th century, fluorspar has been used in smelting iron ores, helping the mixture flow. The Latin word for "flow" is *fluo*, hence the name "fluorspar" and from there, "fluorine".

French chemist Henri Moissan was the first to produce fluorine gas, in 1886. The starting point for most of the varied industrial uses of fluorine – including fluorine gas – is hydrogen fluoride, which is known as hydrofluoric acid when in solution. It is used to make a number of organic (carbon-based) compounds, such as polytetrafluoroethylene (PTFE), whose main applications are as white, waxy, waterproof tape used by plumbers, and, in an "expanded" form, as the breathable fibres in Gore-Tex® jackets. Known by its trade name Teflon®, PTFE is also used as the coating of non-stick cooking pans.

Atomic Number 17
Atomic Weight 35.45
Melting Point −102°C (−151°F)
Boiling Point −34°C (−29°F)

Humphry Davy, realized that the pale green gas chlorine is an element in 1810. He named it after the ancient Greek *chloros*, "pale green". Chlorine is used to manufacture a range of organic chlorine compounds, including chloroform (trichloromethane), mostly used in the production of polytetrafluoroethylene (PTFE). Hydrochloric acid, made by reacting hydrogen and chlorine, is used to make PVC (polyvinyl chloride). Hypochlorite compounds are used in bleaches, water supplies and swimming pools.

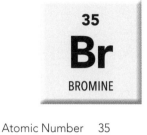

Atomic Number	35
Atomic Weight	79.90
Melting Point	−7°C (19°F)
Boiling Point	59°C (138°F)

Red-brown bromine is one of the two elements (the other is mercury, *see page 94*) that are liquid at room temperature. It forms a pungent vapour – its name comes from the Greek word *bromos*, meaning "stench". French apothecary Antoine-Jérôme Balard was the first to extract it, in 1825. Bromine is produced from seawater. It is used to make flame retardants that are mixed into plastics. Many other uses of bromine have been phased out, because of concerns over the element's toxicity.

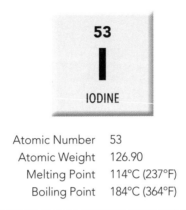

Atomic Number	53
Atomic Weight	126.90
Melting Point	114°C (237°F)
Boiling Point	184°C (364°F)

Iodine is a blue-black solid with a metallic sheen. Even at room temperature, it is surrounded by a deep violet vapour. Iodine was discovered in 1811, by French chemist Bernard Courtois. French chemist Joseph Gay-Lussac named the element, from the Greek *iodes*, meaning "violet". Tincture of iodine is used as an antiseptic and in water purification. Iodine is involved in the synthesis of two thyroid hormones. Good dietary sources include seafood, dairy products and fruit and vegetables.

Atomic Number	85
Atomic Weight	(210)
Melting Point	302°C (576°F)
Boiling Point	335°C (635°F), estimated

Highly radioactive, astatine exists naturally only as a result of the decay of other radioactive elements within uranium ores. The are just a few grams in the entire Earth's crust at any time. It was first identified in 1940, after being created in a laboratory at the University of California. The element's name comes from the Greek word *astatos*, meaning "unstable". The isotope astatine-211, produced in nuclear research facilities, shows promise in the fight against cancer.

CHAPTER

14

THE NOBLE GASES

The elements of Group 18 of the periodic table – the noble gases – are all very inert (unreactive) gases. Radon, the heaviest naturally occurring noble gas, is highly radioactive. All of the naturally occurring elements of Group 18 were either discovered or first isolated by Scottish chemist William Ramsay. This group also includes the transuranium element oganesson.

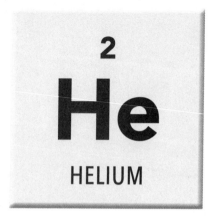

Atomic Number 2
Atomic Weight 4.00
Melting Point −272°C (−457°F),
               only under high pressure
Boiling Point −269°C (−452°F)

THERE IS ALMOST NO HELIUM IN EARTH'S
ATMOSPHERE, SINCE HELIUM ATOMS ESCAPE
EASILY INTO SPACE.

Like all the noble gases, helium is colourless, odourless and extremely unreactive. It has the lowest melting and boiling points of any element. Helium is being made inside every shining star, in nuclear reactions in which hydrogen nuclei join (fuse). It is produced by the decay of radioactive elements underground, from where that helium is extracted.

Helium was discovered in 1868, when French physicist Pierre Jules Janssen and English physicist Joseph Norman Lockyer independently noticed a previously unknown bright line in the spectrum of sunlight. Lockyer suggested the name "helium", from the Greek word *helios*, meaning "sun". Scottish chemist William Ramsay was the first to isolate the gas, in 1895.

Most modern applications of helium involve it being used in liquid form. Liquid helium is used to cool superconducting magnets used for MRI (magnetic resonance imaging), for example. Inhaling helium from party balloons makes your voice sound squeaky. Helium is non-toxic, so this practice is normally safe – but with lungs full of helium, oxygen cannot reach the body's tissues, and a person can suffocate.

10

Ne

NEON

Atomic Number	10
Atomic Weight	20.18
Melting Point	−249°C (−415°F)
Boiling Point	−246°C (−411°F)

NEON MAKES UP ABOUT 0.002 PER CENT
OF THE EARTH'S ATMOSPHERE.

Neon is colourless, odourless and inert. The name is derived from the Greek word *neos*, meaning "new". Neon is the fourth most abundant element in the universe overall, but is very rare on Earth. Scottish chemist William Ramsay and English chemist Morris Travers discovered neon in a sample of liquefied air in 1898. The two chemists separated off the part of the air they knew was argon, and allowed that to warm separately, to see if another element was present. When they passed an electric current through the gas given off, the gas emitted a bright red-orange glow. Analysis of the spectrum of the light confirmed that it was a previously unknown element.

In 1910, French inventor Georges Claude found a way to make use of the bright glow produced when electric current passes through rarefied (low-pressure) neon gas: he invented the neon tube. Claude's invention became commonplace in advertising, since long neon tubes could be shaped into eye-catching product names and slogans. Today's "neon" tubes contain various mixtures of noble gases, and sometimes mercury vapour too. There are many other ways of producing eye-catching lighting, so the use of true neon signs is declining.

18

Ar

ARGON

Atomic Number	18
Atomic Weight	39.95
Melting Point	−189°C (−309°F)
Boiling Point	−185°C (−303°F)

Argon is a colourless and inert gas. Its name comes from the Greek word *argos*, which means "lazy". It accounts for nearly 1 per cent of (dry) air. In 1894, English physicist Lord Rayleigh and Scottish chemist William Ramsay were able to isolate argon from the air. Along with neon, argon is used in "neon" tubes. As a poor conductor of heat, argon is used to fill the gap in double glazing, and it is also used in medical lasers.

36

Kr

KRYPTON

Atomic Number	36
Atomic Weight	83.80
Melting Point	−157°C (−251°F)
Boiling Point	−153°C (−244°F)

Krypton is another inert noble gas, although it can be made to take part in some reactions under extreme circumstances. Scottish chemist William Ramsay, with his assistant Morris Travers, discovered the element, in 1898, by observing previously unknown lines in the spectrum of liquefied air. Since they had not yet isolated it, they based its name on the Greek word *kryptos*, meaning "hidden". Krypton is used in "neon" lights, and krypton-filled arc lamps, installed on airport runways, are visible even through thick fog.

Atomic Number   54
Atomic Weight   131.29
Melting Point   −112°C (−169°F)
Boiling Point   −108°C (−163°F)

Xenon gas is more than 30 times as dense as helium gas; a xenon-filled balloon is heavier than air and plummets to the ground. Like krypton, it can form compounds, albeit under extreme circumstances. Xenon was discovered, in 1898, by Scottish chemist William Ramsay. He based the name on the Greek word *xenos*, meaning "strange". Xenon is used in photography flash units; xenon-filled lamps are also used in strobe lighting and in some high-intensity vehicle headlamps.

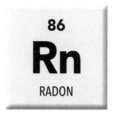

Atomic Number	86
Atomic Weight	(222)
Melting Point	−71°C (−96°F)
Boiling Point	−62°C (−79°F)

Radon gas accounts for more than half the natural background radiation on Earth. It is constantly created as other radioactive elements decay. It collects in underground spaces and can find its way into the air, becoming a health hazard for humans. Radon was first isolated by Scottish chemist William Ramsay, in 1904, after several scientists had detected it as an "emanation" from other radioactive elements, including radium. The element's name is a shortened version of "radium emanation".

THE
TRANSURANIUM
ELEMENTS

Advances in nuclear physics have led to the synthesis of transuranium elements – elements with an atomic number greater than 92, the atomic number of uranium – in nuclear reactors and particle accelerators. All these elements are unstable and radioactive – and only elements up to atomic number 99 have been produced in quantities large enough to see with the naked eye.

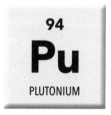

93

Np

NEPTUNIUM

Atomic Weight
(237)

In 1940, American physicists Edwin McMillan and Philip Abelson succeeded in creating element 93, by bombarding uranium with slow neutrons. Named after the planet Neptune, the longest-lived isotope has a half-life of 2.1 million years.

94

Pu

PLUTONIUM

Atomic Weight
(244)

Plutonium was created in 1940 by a team headed by American nuclear chemist Glenn Seaborg, first as the decay product of newly created neptunium, but later by hitting uranium nuclei with deuterons (particles made of one proton and one neutron).

Americium was first created in 1944 by Glenn Seaborg's team, by bombarding plutonium nuclei with alpha particles (helium-4 nuclei) in a cyclotron, an early particle accelerator. The isotope americium-241 is commonly used in smoke detectors.

Atomic Weight
(243)

Curium was also created in 1944, in the same way and by the same team as americium – but three months earlier. It was named after Polish physicist and chemist Marie Curie and her husband, Pierre.

Atomic Weight
(247)

97

Bk

BERKELIUM

Atomic Weight
(247)

Berkelium was named after the University of California, Berkeley, where it was produced, in 1949, by Glenn Seaborg's team, who bombarded americium nuclei with alpha particles in a cyclotron. Berkelium has no practical applications outside scientific research.

98

Cf

CALIFORNIUM

Atomic Weight
(251)

Californium was produced by the same team as berkelium, by bombarding curium nuclei with alpha particles. The newly discovered element was named after the American state in which the experiment took place. Californium-252 is used to treat some cancers.

Together with fermium, einsteinium was first detected in the fallout of the first ever test of a hydrogen bomb, in 1952. Both had been produced after uranium nuclei had absorbed several neutrons. Einsteinium was named after German-born physicist Albert Einstein.

99

Es

EINSTEINIUM

Atomic Weight
(252)

Together with einsteinium, fermium was first detected in the fallout of the first ever test of a hydrogen bomb, in 1952. Both had been produced after uranium nuclei had absorbed several neutrons. Fermium was named after Italian physicist Enrico Fermi.

100

Fm

FERMIUM

Atomic Weight
(257)

101

Md

MENDELEVIUM

Atomic Weight
(258)

A team led by Glenn Seaborg and American nuclear scientist Albert Ghiorso discovered mendelevium in 1955, in the cyclotron at the University of California, Berkeley, by bombarding einsteinium with alpha particles. The element is named after Russian chemist Dmitry Mendeleev.

102

No

NOBELIUM

Atomic Weight
(259)

In 1956, a team at the Joint Institute for Nuclear Research in Dubna, Russia, produced element 102. In 1957, a team at the Nobel Institute of Physics, in Sweden, erroneously claimed that they had produced element 102 – and they suggested the name nobelium, after Alfred Nobel.

Element 103, lawrencium, was first produced in 1961, by a team led by Albert Ghiorso at Berkeley, by bombarding californium with ions of the element boron. It is named after the inventor of the cyclotron, American physicist Ernest Lawrence.

Atomic Weight
(262)

There was controversy over the first claim of the discovery of element 104, in 1964, although it has been studied in detail since. The element was named, in 1997, after New Zealand-born physicist Ernest Rutherford.

Atomic Weight
(267)

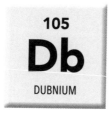

105

Db

DUBNIUM

Atomic Weight
(268)

There was controversy over the first claim of the discovery of element 105, in 1967, although it has been studied in detail since. The element was named, in 1997, after Dubna, Russia, where the first claim was made.

106

Sg

SEABORGIUM

Atomic Weight
(269)

There was controversy over the first claim of the discovery of element 106, in 1974, although it has been studied in detail since. The element was named, in 1997, after American nuclear chemist Glenn Seaborg.

The first convincing evidence for element 107 came from a team, led by Peter Armbruster, at the Gesellschaft für Schwerionenforschung (GSI, the Society for Heavy Ion Research) in Darmstadt, Germany, in 1981. The element is named after Danish physicist Niels Bohr.

Atomic Weight
(270)

Peter Armbruster's team at the GSI (*see* bohrium, *above*) were also first to produce element 108, which was named after the word *Hassia*, Latin for the German state of Hesse where the laboratory is located.

Atomic Weight
(269)

109

Mt

MEITNERIUM

Atomic Weight
(278)

Element 109 was also discovered by Peter Armbruster's team at the GSI (*see* bohrium, *page 185*). The element was named after German physicist Lise Meitner.

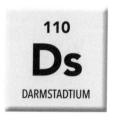

110

Ds

DARMSTADTIUM

Atomic Weight
(281)

Various isotopes of element 110, darmstadtium, were produced at both the GSI (*see* bohrium, *page 185*) and Dubna (*see* nobelium, *page 182*) from 1987 onwards. The element's name is derived from the city of Darmstadt, where the GSI is located.

Element 111 was first produced at the GSI (*see* bohrium, *page 185*), in 1994. It is named after German physicist Wilhelm Conrad Röntgen, known for his pioneering work with X-rays.

Atomic Weight
(281)

Element 112, copernicium, was first created at the GSI (*see* bohrium, *page 185*), in 1996. It is named after the 16th-century Polish astronomer Nicolaus Copernicus.

Atomic Weight
(285)

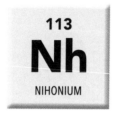

Atomic Weight
(286)

Element 113 was first produced by a team of scientists at Rikagaku Kenkyūjo (RIKEN) in Wakō, Japan. The name, approved by IUPAC (the International Union of Pure and Applied Chemistry) comes from the common Japanese name for Japan (*nihon*).

Atomic Weight
(289)

Element 114 was first created in 1998, at the Flerov Laboratory of Nuclear Reactions at the Joint Institute for Nuclear Research (JINR) in Dubna, Russia, where scientists bombarded plutonium-244 nuclei with nuclei of calcium-48.

Element 115 was created at the JINR (*see* flerovium, *page 188*), in 2003. IUPAC (*see* nihonium, *page 188*) confirmed its name, derived from the name of capital of Russia, in 2016. Only about 100 atoms of moscovium have been observed.

Atomic Weight
(289)

Scientists at the Lawrence Livermore National Laboratory in the United States collaborating with the JINR in Russia discovered livermorium in the early 2000s. The longest-lived isotope, livermorium-293, has a half-life of about 60 milliseconds.

Atomic Weight
(293)

117

Ts

TENNESSINE

Atomic Weight
(294)

Element 117 was produced in 2009 and its existence was announced in 2010, after a collaboration between American and Russian scientists. Tennessine, named after the US state of Tennessee, has the suffix -ine, like all Group 17 elements.

118

Og

OGANESSON

Atomic Weight
(294)

Oganesson was made in 2002 by Russian and American scientists, at the JINR (*see flerovium, page 188*). It was named after the head of the JINR team, Yuri Oganessian. Only five or six atoms of oganesson have been observed.

INDEX OF ELEMENTS